Pre-Calculus
Math Tutor Lesson Plan Series

iGlobal Educational Services

To order, contact iGlobal Educational Services
PO Box 94224, Phoenix, AZ 85070
Website: www.iglobaleducation.com
Fax: 512-233-5389

©2017 by iGlobal Educational Services, Austin, Texas.

ISBN-13: 978-1-944346-61-4

Printed in the United States of America.

Pre-Calculas

Contents

Introduction

Tutoring is beginning to get the respect and recognition it deserves. More and more learners require individualized or small group instruction whether it is in the classroom setting or in a private tutoring setting either face-to-face or online.

This lesson plan book is the sixth out of sixth titles in the series "Math Tutor Lesson Plan" Series. It is conceived and created for tutors and educators who desire to provide effective tutoring either in person or online in any educational setting, including the classroom.

Inside This Lesson Plan Book

This *Pre-Calculus: Math Tutor Lesson Plan Series* book provides appropriate practice during tutoring sessions for learners for both face-to-face and online tutoring sessions focused on topics in Pre-Calculus.

The goal of the *Pre-Calculus: Math Tutor Lesson Plan Series* book is to support all types of tutors. Also, this book is to support teachers who want to provide in-class tutoring to their students in either an individualized or small group tutoring setting. Lastly, this book is also for teachers who are providing math intervention either individually or in small group tutoring sessions either face to face or online so that they can select the specific lesson plan to address the learner's math learning needs.

How to Use This Lesson Plan Book

iGlobal Educational Services, in collaboration with, Dr. Alicia Holland-Johnson, Tutor Expert and Consultant, created this tutoring resource to help with designing effective tutoring instruction for tutors and teachers who desire to provide in-class tutoring sessions.

These specific lessons were selected based upon field-tested experiences with learners who had learning needs over the years in these specific areas in mathematics. We have provided learning objectives and specific topics covered in each tutoring session so that you can align them with your state's specific standards or adapted standards. For overseas tutors, you can follow suite and align the lesson objectives to specific educational standards required in your country.

These lesson plans should be used to supplement a strong and viable curriculum that encourages differentiation for all diverse learners. They can be used in individual or small group tutoring sessions conducted face-to-face or online in any educational setting, including the classroom.

Organization of the Lesson Plan Book

Rather than provide a specific "curriculum" to follow, *Pre-Calculus: Math Tutor Lesson Plan Series* book provides a blueprint to design effective tutoring lessons that are aligned with the *"Dr. Holland-Johnson's Session Review Framework"*. Tutor evaluators and coaches are able toanalyze tutoring sessions and coach tutors when utilizing the *"Dr. Holland-Johnson's Lesson Plan Blueprint for Tutors"*. In each lesson plan, learners have an opportunity to focus on real-world connections, vocabulary, and practice the math concepts learned in the tutoring sessions in the appropriate amounts to learn and retain the content knowledge. Tutors will have an opportunity to provide direct and guided instruction, while learners practice concepts on their own during independent instruction.

Each lesson plan comes with a mini-assessment pertaining to the math concepts learned in the specific tutoring session. Depending on the learner's academic needs, the tutor or teacher will deem when it is appropriate to administer the mini-assessment. For online tutoring sessions or as an online option to take the mini-assessment, tutors and teachers can upload these mini-assessments to be completed online in their choice of an online assessment tool.

Lesson 1
Limit of Functions

Lesson Description:

This lesson is designed to help students identify where functions are discontinuous and evaluate limits of functions. Additionally, students will have an opportunity to learn how to distinguish between convergent and divergent sequences.

Please be sure to utilize the questions to help spark student engagement and cover the vocabulary that is associated with this specific tutoring session. For your own knowledge, sample responses have been provided to guide you as well.

Learning Objectives:

In today's lesson, the learner will learn the limit of functions with 75% accuracy in 3 out of 4 trials.

Introduction

Consider a situation where a ball bounces back and forth, and back and forth. It continues to do so uncounted times. Likewise, a vibrating string or a wire that vibrates with large amplitudes (distance from central points) continues to vibrate continuously while gradually reducing their size. The expected amplitude of this string after uncounted times tends to be 0. This is a notion of limits. That is, a limit of the amplitude of a vibrating string is zero as more and more vibrations are made. The concept of limits is highly applicable in real life situation.

Questions to Spark Student Engagement

- Have you encountered a function that does not exist at a certain point? How
- would you describe that function?
- From your current knowledge about functions, what do you think are continuous functions?

Real-World Connections

There are many real-world connections for this particular lesson. Please make sure that you communicate this clearly to your learner prior to beginning the tutoring session.

- Knowing how to evaluate limits will help in calculating the rates of change as applied to capacity, speed, economics and other real-life situations
- Understanding the continuity of a function assists in better identifying its domain and range
- Determining the convergence of a sequence helps in identifying the upper or lower limits of infinitely happening scenarios, if any

Specific Vocabulary Covered

The learner needs to know these vocabulary terms by the end of the session. As a suggestion, you can have him or her write them on flashcards or even use them as visual vocabulary words.

Limit
The limit of the function f(x) is the number, L, that the function approaches as x approaches a number c. We say the limit of f(x) as x approaches c, equals to L and write

$$\lim_{x \to c} f(x) = L,$$

Left-hand Limit
The left-hand limit of the function f(x) is the number, L, that the function approaches as x approaches a number c from the left, that is x<c and we write

$$\lim_{x \to c^-} f(x) = L$$

Right-hand Limit
The right-hand limit of the function f(x) is the number, L, that the function approaches as x approaches a number c from the right, that is, x>c and we write

$$\lim_{x \to c^+} f(x) = L$$

Continuous Function
A function f(x) is continuous at a point x=c if the function f(x) is defined at c and

$$\lim_{x \to c} f(x) = f(c)$$

Limit to Infinity
If f is a function defined on the interval (c, ∞), the limit to infinity is the value f(x) approaches as x approaches infinity.

Convergent Sequences
A sequence $\{a_n\}$ is convergent if its limit as n approaches infinity exists, that is $\lim_{n \to \infty} a_n$ exists.

Divergent Sequences
A sequence $\{a_n\}$ is divergent if its limit as n approaches infinity does not exists, that is If $\lim_{n \to \infty} a_n$ does not exist.

What is Limits?

- Let $f(x) = 2x$

- How do you describe the behavior of the graph of the equation (shown on the right) as x approaches 1?

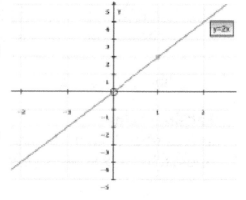

- What happens to the graph for x values approaching 1 from the right?

- How about for values approaching from the left?

- As x values from the left approach 1, the corresponding y values approach 2

- As x values from the right approach 1, the corresponding y values also approach 2

- In general, for any function f(x) with a point (c,L), the limit of the function is written as follows:

(read as: "the limit of f(x) as x approaches c is equal to L")

Example:

Let $f(x) = x + 2$. Find $\lim\limits_{x \to 3} f(x)$

There are different ways to solve for this:

Through tables – values from the left and right approach the same number

x	2.5	2.9	2.995	3	3.005	3.1	3.5
y	4.5	4.9	4.995	5	5.005	5.1	5.5

Through direct substitution

$$f(x) = x + 2$$
$$f(3) = 3 + 2 = 5$$

Through graph

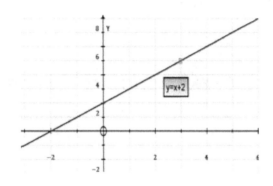

More on Limits

Now, let's look at this function: $f(x) = \dfrac{|x+2|}{x+2}$

What do you think is $\lim\limits_{x \to -2} f(x)$?

Applying direct substitution is not possible as f(x) does not exist at x = -2

If the function does not exist at a certain point, does it also mean that the limit of the function as x approaches that point does not exist as well?

Let's look at the graph and see how the function behaves from either side of x = -2

As *x* approaches -2 from the <u>left</u>, *y* approaches -1; this is called the **left-hand limit**

$$\lim_{x \to -2^-} f(x)$$

As *x* approaches -2 from the <u>right</u>, *y* approaches 1; this is called the **right-hand limit**

$$\lim_{x \to -2^+} f(x)$$

$$\lim_{x \to -2^-} f(x) = \lim_{x \to -2^+} f(x)$$

For a limit to exist, and each limit must also exist.

Properties of Limits

Given that: $\lim\limits_{x \to c} f(x) = L$ and $\lim\limits_{x \to c} g(x) = M$

and L and M are real numbers,

1. Scalar multiple $\qquad \lim\limits_{x \to c} \left[af(x) \right] = aL$

2. Sum or difference $\qquad \lim\limits_{x \to c} \left[f(x) \pm g(x) \right] = L \pm M$

3. Product $\qquad \lim\limits_{x \to c} \left[f(x) g(x) \right] = LM$

4. Quotient $\qquad \lim\limits_{x \to c} \dfrac{f(x)}{g(x)} = \dfrac{L}{M}$, provided M = 0

5. Power $\qquad \lim\limits_{x \to c} \left[f(x) \right]^n = L^n$

For polynomial and rational functions, the following properties apply:

1. Given a polynomial function f(x) and a real number c, then

$$\lim\limits_{x \to c} f(x) = f(c)$$

2. Given a rational function $r(x) = \dfrac{p(x)}{q(x)}$, and a real number c such that $q(c) \neq 0$, then

$$\lim\limits_{x \to c} r(x) = r(c) = \dfrac{p(c)}{q(c)}$$

Evaluating Limits

The following examples show how limits are evaluated using the properties

1. $\lim\limits_{x \to 2}\left(x^2 - 5x - 2\right)$

   What kind of function is this? What property can be used for this?

   $$\lim\limits_{x \to 2}\left(x^2 - 5x - 2\right) = \left[(2)^2 - 5(2) - 2\right] = 4 - 10 - 2 = -8$$

2. $\lim\limits_{x \to 3}\dfrac{2x}{2x - 1}$

   What kind of function is this? What property can be used for this?

   $$\lim\limits_{x \to 3}\dfrac{2x}{2x - 1} = \dfrac{2(2)}{2(2) - 1} = \dfrac{4}{4 - 1} = \dfrac{4}{3}$$

Evaluating Limits – Indeterminate Form

What if $\lim\limits_{x \to c} f(x) = 0$ and $\lim\limits_{x \to c} g(x) = 0$, what do you make of $\lim\limits_{x \to c}\dfrac{f(x)}{g(x)}$?

This case will result in $\dfrac{0}{0}$ or known as **indeterminate form**.

For indeterminate forms, the functions need to be simplified first through factoring and dividing, and rationalizing.

Here's an example:

1. $\lim\limits_{x \to 2}\dfrac{2x^2 - 3x - 2}{x^2 + x - 6}$ What happens if direct substitution is used?

   indeterminate form

   $$\lim\limits_{x \to 2}\dfrac{2x^2 - 3x - 2}{x^2 + x - 6} = \dfrac{2(2)^2 - 3(2) - 2}{(2)^2 + 2 - 6} = \dfrac{8 - 6 - 2}{4 + 2 - 6} = \dfrac{0}{0}$$

So we need to factor out and divide the polynomials

$$\lim_{x \to 2} \frac{2x^2 - 3x - 2}{x^2 + x - 6} = \frac{(2x+1)(x-2)}{(x+3)(x-2)} = \frac{(2x+1)}{(x+3)}$$

Now, we can do direct substitution to the simplified form

$$\lim_{x \to 2} \frac{2x+1}{x+3} = \frac{2(2)+1}{2+3} = \frac{5}{5} = 1$$

Continuity of Functions

A function is said to be continuous when the left-hand limit and the right-hand limit exist and are equal, and f(c) = L.

$$\lim_{x \to c-} f(x) = \lim_{x \to c+} f(x) = f(c) = L$$

Example:

1. $f(x) = \dfrac{x}{(x+2)(x-3)}$ Is this function continuous on all values of x?

- To solve for this, we first need to get the "critical points" on this function. Analyzing, the two critical points are -2 and 3 as these are the numbers that will make the denominator 0.

- Since the function does not exist at these points, the function is said to be continuous on all values of x, except x = -2 and x = 3.

Limits at Infinity

To evaluate limits at infinity, the following information is used:

$$\lim_{x \to -\infty} \frac{k}{x^r} = 0 \text{ , where k and r are real numbers not equal to 0}$$

$$\lim_{x \to +\infty} \frac{k}{x^r} = 0$$

Example:

$$f(x) = \frac{4x^3 - 5x + 3}{2x^3 - x + 5} \qquad \text{Find} \quad \lim_{x \to \infty} f(x)$$

- First thing to do here is to identify the highest exponent (degree) of the function

- Multiply both the numerator and denominator by $\frac{1}{x^n}$, where n is the highest exponent

- Apply the concept learned on evaluating limits at infinity

$$\lim_{x \to \infty} \frac{4 - \dfrac{5x}{x^3} + \dfrac{3}{x^3}}{2 - \dfrac{x}{x^3} + \dfrac{5}{x^3}} = \lim_{x \to \infty} \frac{4 - \dfrac{5}{x^2} + \dfrac{3}{x^3}}{2 - \dfrac{1}{x^2} + \dfrac{5}{x^3}} = \frac{4 - 0 + 0}{2 - 0 + 0} = \frac{4}{2} = 2$$

- Limits at infinity is used in sequences to identify whether they are convergent or divergent

- For a sequence U_n, such that f(n) – U_n for every positive integer n, then

$$\lim_{n \to \infty} U_n = L$$

- That means the sequence converges to L, hence, a convergent sequence

- If $\lim_{n \to \infty} U_n$ does not exist, then the sequence is call divergent sequence

Question 1:

Determine where this function is continuous $f(x) = \dfrac{x-1}{4x^2-9}$

Solution:

The function $f(x)$ is discontinuousxwhen $4x^2-9=0$.

$$4x^2 = 9 \quad ; \quad x^2 = \frac{9}{4}$$

Taking square roots on both sides, we get $x = \sqrt{\dfrac{9}{4}} = \pm\dfrac{3}{2} = \pm 1.5$

Thus, $f(x)$ is discontinuous at $x = \dfrac{2}{3} = 1.5$ and $x = -\dfrac{3}{2} = 1.5$

Teacher Questions:

- **What value(s) of x will you use to identify the limit?**

  Values close but not equal to $x = \dfrac{2}{3} = 1.5$ and $x = -\dfrac{3}{2} = 1.5$

- **Does the left-hand limit for the value(s) of x exist?**

  $\displaystyle\lim_{x \to 1.5^-} \frac{x-1}{4x^2-9} = -\infty$ and $\displaystyle\lim_{x \to -1.5^-} \frac{x-1}{4x^2-9} = \infty$

  Therefore, the left hand limits do not exists

- **Does the right-hand limit for the value(s) of x exist?**

  $\displaystyle\lim_{x \to 1.5^+} \frac{x-1}{4x^2-9} = \infty$ and $\displaystyle\lim_{x \to -1.5^+} \frac{x-1}{4x^2-9} = -\infty$

- **Is f(x) for the identified value(s) of x equal to the limit?**

  No

Question 2:

Is the sequence $U_n = \frac{3n+2}{n-1}$ convergent or divergent?

Solution:

We determine the following limit $\lim\limits_{n\to\infty} \frac{3n+2}{n-1}$

We divide through by n

$$\lim\limits_{n\to\infty} \frac{3n+2}{n-1} = \lim\limits_{n\to\infty} \frac{3+\frac{2}{n}}{1-\frac{1}{n}} = \frac{3+0}{1-0} = \frac{3}{1} = 3$$

Thus, the limit approaches a specific point showing that it is convergent

Teacher Questions:

- **How will you approach this problem?**
  Use infinity limit approach so determining the limit of the function

- **Will getting the limit be useful in this problem?**
  Yes

- **If so, what kind of limit will you use?**
  Limit at infinity

- **What does the result say about the sequence?**
  The sequence is convergent

Lesson Reflection:

In this lesson, we have learned about limits of functions. As we close our lesson today, please reflect on the following questions:

- How would you distinguish convergent and divergent sequences?

- What was your biggest take-a-way about evaluating limits of functions?

Video Suggestions

While there are many videos available to help you, these are only to be used as a starting point to help you with any supplemental videos in which you may use. Please conduct a search on either YouTube or Teacher Tube to find appropriate videos for this lesson.

- Tricks to finding limits:

- Limits: A Visual Approach

- Continuity and limits

Question 1:

Find the limit, if it exists: $\lim\limits_{x \to -3} \dfrac{x^2+x-6}{x+3}$

Explanation:

Since direct substitution will result in an indeterminate form, the numerator will be factored out instead and simplified

$$\lim_{x \to -3} \frac{x^2+x-6}{x+3} = \frac{(x+3)(x-2)}{x+3} = x-2$$

Now, direct substitution should work

$$\lim_{x \to -3}(x-2) = -3-2 = -5$$

Question 2:

Find the limit, if it exists: $\lim\limits_{x \to \infty} \dfrac{2x^4+3x^3-6}{x^4+2x^2+10}$

Explanation:

Limit at infinity requires simplifying the equation to be able to use the concept

$$\lim_{x \to \infty} \frac{2x^4+3x^3-6}{x^4+2x^2+10} = \frac{\dfrac{2x^4}{x^4}+\dfrac{3x^3}{x^4}-\dfrac{6}{x^4}}{\dfrac{x^4}{x^4}+\dfrac{2x^2}{x^4}+\dfrac{10}{x^4}} = \frac{2+\dfrac{3}{x}-\dfrac{6}{x^4}}{1+\dfrac{2}{x}+\dfrac{10}{x^4}}$$

Applying the concept of rules for limits at infinity

$$\lim_{x \to \infty} \frac{2+\dfrac{3}{x}-\dfrac{6}{x^4}}{1+\dfrac{2}{x}+\dfrac{10}{x^4}} = \frac{2+0-0}{1+0+0} = 2$$

Question 3:

Identify where this function is continuous: $f(x) = \sqrt{x-9}$

Explanation:

- Since this is a radical equation, the rules of radicals will be followed

- For this equation to exist, $x \geq 9$. Any other values less than 9 are not possible.

- Hence, the function is continuous only when $x \geq 9$

Mini-Assessment

At the end of your tutoring session, these are the questions that the learner will need to complete. Please make sure that you keep in mind only the topics in which you have covered in the lesson. As a suggestion, if time does not permit, you can have the learner complete the rest of the mini-assessment at the next tutoring session.

1. When is a function said to have a limit at a point?

    A. When the sided limits exists
    B. When at least one sides limit exists
    C. When both sided limits do not exists
    D. When both sided limits exists and are equal
    E. None of the above

2. Find the limit of the functions f(x)=3x+3 as x approached 0.

    A. 1   B. Does not exists   C. 0   D. 3   E. 6

3. Does the limit of the following function as x approaches 1 exists?

$$f(x) = \frac{x+1}{x^2+1}$$

    A. No, the sided limits are not the same
    B. Yes, the limit is zero
    C. Yes, the limit is 0.5
    D. No, both sided limits approach infinity
    E. None of the above

4. Determine the sided limit

$$\lim_{x \to 1^-} \frac{1}{1-x}$$

    A. 0
    B. $-\infty$
    C. 0.5
    D. 1
    E. ∞

5. Evaluate $\lim_{x \to -\frac{1}{2}} e^{2x+1}$

A. 0
B. e
C. 1
D. -1
E. Undefined

6. Determine the following limit

$$\lim_{x \to -2} \frac{x^2 + 5x + 6}{2x + 4}$$

A. 0
B. 1/2
C. 5/2
D. Does not exists
E. -1/2

7. Evaluate the limit $\lim_{x \to 0} f(x)$ if

$$f(x) = \begin{cases} 2 & x \le 0 \\ \dfrac{x}{3} + 1 & x > 0 \end{cases}$$

A. 2
B. Does not exists
C. 3
D. 1
E. 0

Mini-Assessment Answers and Explanations

1. Answer D

Explanation:
The limit of a function exists if both sided limits exists and are equal

2. Answer D

Explanation:
Since the function is s polynomial, we substitute for x.

$$\lim_{x \to 0} 3x + 3 = 3(0) + 3 = 3$$

3. Answer C

Explanation:

$$\lim_{x \to 1} \frac{x - 1}{x^2 - 1} = \lim_{x \to 1} \frac{x - 1}{(x - 1)(x + 1)} = \lim_{x \to 1} \frac{x - 1}{(x - 1)(x + 1)}$$
$$= \lim_{x \to 1} \frac{1}{(x + 1)}$$
$$= \frac{1}{2} = 0.5$$

4. Answer E

Explanation:
In the problem, x approached 1 from the lower sides, hence we substitute values less than 1. This implies, *1-x>0* for all *x<-1*.

Thus, the limit will be a positive number.

Since 1/(1-1) is undefined, the limit is ∞.

5. Answer C

Explanation:
Since -½ is in the domain of the function, we simply do substitution

$$\lim_{x \to -\frac{1}{2}} e^{2x+1} = e^{2(-\frac{1}{2})+1} = e^0 = 1$$

6. Answer B

Explanation:

$$\lim_{x \to -2} \frac{x^2 + 5x + 6}{2x + 4} = \lim_{x \to -2} \frac{x^2 + 3x + 2x + 6}{2x + 4} = \lim_{x \to -2} \frac{x(x + 3) + 2(x + 3)}{2x + 4}$$

$$\lim_{x \to -2} \frac{(x + 2)(x + 3)}{2(x + 2)} = \lim_{x \to -2} \frac{(x + 3)}{2}$$

$$= \frac{-2 + 3}{2} = \frac{1}{2}$$

7. Answer B

Explanation:
We determine the sided limits

$$f(x) = \begin{cases} 2 & x \leq 0 \\ \dfrac{x}{3} + 1 & x > 0 \end{cases}$$

$$\lim_{x \to 0^-} f(x) = \lim_{x \to 0^-} 2 = 2$$
$$\lim_{x \to 0^+} f(x) = \lim_{x \to 0^-} \frac{x}{3} + 1 = \frac{0}{3} + 1 = 1$$

Since the two sided limits are not equal, the limit of the function as x approaches zero does not exists.

Lesson Reflection:

In this lesson, we have learned about limits of functions. As we close our lesson today, please reflect on the following questions:

■ How would you distinguish convergent and divergent sequences?

■ What was your biggest take-a-way about evaluating limits of functions?

Lesson 2
Parametric & Polar Equations

Lesson Description:

This lesson is designed to help students describe phenomena when using both parametric and polar equations. Additionally, students will have an opportunity to model and solve problems that will require parametric and polar equations. Please be sure to utilize the questions to help spark student engagement and cover the vocabulary that is associated with this specific tutoring session. For your own knowledge, sample responses have been provided to guide you as well.

Learning Objectives:
In today's lesson, the learner will learn parametric equations and polar equations with 75% accuracy in 3 out of 4 trials.

Introduction

Now, it's the learner's turn. This is the learner's chance to demonstrate that he or she can complete the skill on his or her own. Depending on how much time in which you have left in the session, you may want to use only one or two of these questions and focus on the mini-assessment questions that are aligned with this particular lesson. Of course, it will depend on both the session time and the learner's progress.

Questions to Spark Student Engagement

- Have you tried graphing the following equation?
- How do you setup equations of motions of objects as functions of time?

Real-World Connections

There are many real-world connections for this particular lesson. Please make sure that you communicate this clearly to your learner prior to beginning the tutoring session.

■ Understanding how parametric equations are used will help in modeling conic functions
■ A certain phenomena can be explained better with parametric and polar equations
■ Real-life motion problems can be solved by using parametric equations

Specific Vocabulary Covered

The learner needs to know these vocabulary terms by the end of the session. As a suggestion, you can have him or her write them on flashcards or even use them as visual vocabulary words.

■ **Parameter**
This is the variable in which we can express the other variables in terms of.

■ **Parameter equation**
This is a set of functions expressed in terms of another variable called the parameter

■ **Polar coordinates**
This is a coordinate system where points are described in terms of length and angles. The angle is measured from the positive *x-direction* while the length is measured from the origin.

■ **Polar equation**
These are equations that are expressed in terms of r and θ where r is the length and θ is an angle.

Parametric Equations

What if you're asked to do graph $\frac{x^2}{a} + \frac{y^2}{b} = 25$ manually? How are you going to do it?

- First of all, the above equation is not a function, it's a relations
- Trying to graph it manually will need you to isolate y and write it in terms of the x-variable; the above equation will be too ugly if you do that

Instead, it is best to express both x and y as dependent variables with the independent variable being t (time); in other words, x and y will both be functions of time, t

$$X = x\,(t) \text{ and } Y = y\,(t)$$

Where t is the parameter, and x(t) and y(t) are the parametric equations

An example of parametric equations are as follows

Setting up the table of values and plotting them on the graph:

t	x	y
0	5	8
1	9	6
2	13	4
3	17	2
4	21	0
5	25	-2

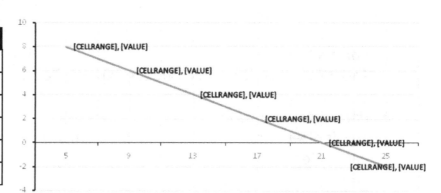

These parametric equations model a uniform linear motion
x(t) = a + bt
y(t) = c + dt

(a, c) are initial positions, b and d are horizontal and vertical velocities, respectively.

Polar Equations

Now that we have a model for linear motion, how do you think are circular motions, and even projectile motions, modeled?

- Before presenting the model for other types of motion, let's first talk about polar equations. Certain functions, particularly circular functions, are best represented in polar coordinate form
- In a polar coordinate system, the equivalent of x and y coordinates are the **radial (r)** and **angular (0)**

Cartesian coordinates as defined in terms of polar coordinates:

$$x = r \cos \theta$$
$$y = r \sin \theta$$

Polar coordinates as defined in terms of Cartesian coordinates:

$$r = \sqrt{x^2 + y^2}$$
$$\theta = \tan^{-1} \frac{y}{x}$$

An example of a polar equation is as follows:

$$r = 2 \cos \theta$$

Setting up the table of values and plotting them on the polar coordinate graph:

Parametric Equations – Non-Linear Motions

Non-linear motions can be modeled using parametric equations and polar equations combined.

For a uniform circular motion:

$$x(t) = r \cos(\theta_0 + \omega t)$$
$$y(t) = r \sin(\theta_0 + \omega t)$$

Where r is the radius of the circle, 0_0 initial angle of the initial position, and w is the angular speed

For a projectile motion:

$$x(t) = x_0 + v_0 \cos(\theta_0 t)$$
$$y(t) = y_0 + v_0 \sin(\theta_0 t) - \frac{g}{2}t^2$$

Where v_0 is the initial velocity, θ_0 initial angle at which the projectile is fired and g is the acceleration due to gravity

Guided Instruction

For this session segment, you will be working with the learner together. This is your chance to act as a guide for the learner and then allow him or her to use both their critical and creative thinking skills. In light of this, there are four teacher questions to help you with this process. Please be sure to use the teacher question answers to double check your work.

Question 1:

Aliya is traveling in a straight line at a constant speed. At t = 5 seconds, she is at position (150,200). At t = 10 seconds, she is found at the point (120,240). What are the parametric equations of Aliya's x and y positions (in feet) as a function of time (seconds)?

- Are we given initial values here? If so, what are they? If none, what are you going to use for the initial values?
- How do you compute for the horizontal and vertical velocities of Aliya?

Solution:

x(t)=-6t+180
y(t)=8t+160

At initial values, let the values be (a_1, a_2)
At *t=5, x=150* and *y=200*.
At *x=10, x=120* and *y=240*

Thus, considering movement in *x* direction, we have coordinates *(5,150)*and *(10,120)*.

The slope is $m = \dfrac{240-200}{10-5} = \dfrac{40}{5} = 8$

While there are many videos available to help you, these are only to be used as a starting point to help you with any supplemental videos in which you may use. Please note that these video URLs were working at the time of publication.

The equation will be $\frac{y-240}{t-10} = 8;\quad y - 240 = 8t - 80\,;\quad y = 8t + 160$

We have *y(t)=8t+160*

Are we given the initial values here? If so, what are they? If none, what are you going to use for the initial values?

We are not given the initial values. We use the values given to determine the initial values

How do you compute for the horizontal and vertical velocities of Aliya?

We take the points (150,200) and (120,240) then calculate the change in displacement in each case and divide by the change in time

Change in time is *(10s-5s)=5s.*

Horizontal velocity = $\frac{120-150}{10-5} = -6\,ft/s$

Vertical velocity = $\frac{240-200}{10-5} = 8\,ft/s$

Question 2:

Car A is traveling at 60 mph and has an initial position direction of 1100, while a second car, Car B, is traveling at 5 mph with an initial position direction of 400. Both cars start at the origin at t = 0. What are the cars' parametric equations for x and y positions (in miles) as a function of time (in hours)? How far apart are the two cars after 3 hours?

- What type of parametric equation will be used to model this scenario?
- How will you setup the equation for the parametric equations?

Solution:

Two cars have horizontal movement therefore; there is change in x direction in both cases.

There is no vertical movement therefore, $y(t)=0$ in both case.

Since Distance = speed × time

The horizontal distance for car A is given by $x=60×t+1100$, since the initial distance is 1100, thus, the respective parametric equation is $x(t)=60t+1100, t≥0$

The parametric equations for car A is $y=0$ and $x=60t+1100, t≥0$

The speed of car B is 5mph and Since Distance = speed × time

The horizontal distance for car B is given by $x=5 × time$

Thus the parametric equation is $x(t)=5t+400$ since the initial distance is 400.

The parametric equations for car B is $y(t)=0$ and $x(t)=5t+400, t≥0$

After three hours car A is at $x=60(3)+1100=1280m$

After three hours car B is at $x=5(3)+400=415m$

Thus, $1280-415=865$ miles is the distance apart after 3 hours.

Question 2:

Best friends Kim and Jai walk to school from their respective homes. On Monday, both started walking at the same time. Kim's house is at (20 , -7) and he walks in the direction 1300 at a speed of 2 m/s. Jai has walked 3 seconds before reaching the point (26 , 10). His walking speed is 2.5 m/s westward direction. Will they meet each other somewhere along the way?

Explanation:

Let us plot the points of Kim's house, the point (26,10) and their movements in a plane.

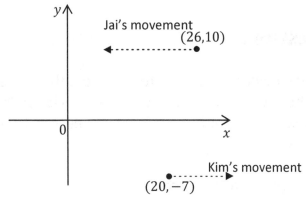

The two friends are walking in parallel directions; therefore, they will not meet

- **What type of parametric equation will be used to model this scenario?**
  Linear parametric equations

- **How will you setup the equation for the parametric equations?**
  Use the definition of linear equations to set it up

  Total distance travelled $x = speed(t) + initial\ distance$

  There is no vertical movement and therefore each car will have a parametric equation $y(t) = 0$.

Video Suggestions

While there are many videos available to help you, these are only to be used as a starting point to help you with any supplemental videos in which you may use. Please conduct a search on either YouTube or Teacher Tube to find appropriate videos for this lesson.

Below are some suggested title searches:

- Parametric Equations

- Modeling motion

- Polar coordinates

Question 1:

Assuming that the home plate is at the origin, John hits the ball at t = 0.5 seconds, 0.75 meters above the ground. He hit it at an angle of inclination of 200 and speed of 50 meters per second. What are the parametric equations for the x and y position of the ball as a function of time?

Explanation:

This is a projectile motion so we are going to use the following equations:

$$x(t) = x_0 + v_0 \cos(\theta_0 t)$$
$$y(t) = y_0 + v_0 \sin(\theta_0 t) - \frac{g}{2} t^2$$

Given:

$X_0 = 0$ $\qquad\qquad$ $y_0 = 0.75$
$V_0 = 50 \text{ m/s}$ $\qquad$ $g = 9.81 \text{ m/s2}$
Initial time $= t - 0.5 \text{ s}$ $\qquad$ $\theta_0 = 200$

Plugging in the values:

$$x(t) = 0 + 50\cos(20°(t-5)) = 50\,cos(20°(t-5))$$
$$y(t) = 0.75 + 50\,sin(20°(t-5)) - \frac{9.81}{2}(t-5)^2$$

Mini-Assessment

At the end of your tutoring session, these are the questions that the learner will need to complete. Please make sure that you keep in mind only the topics in which you have covered in the lesson. As a suggestion, if time does not permit, you can have the learner complete the rest of the mini-assessment at the next tutoring session.

1. Identify the parameter(s) in the following equations

$$x=2t+\cos t$$
$$y=\sin t+1$$

    A. x
    B. t
    C. y and x
    D. x+t
    E. y

2. Write the parametric equations of the function $f(x)=2x^2+4x+1$ **if** $x/_r=2$ **where r is the parameter.**

    A. $f(r)=2r^2+4r+1; \quad x(r)=2r$
    B. $f(r)=4r^2+8r+1; x(r)=r/_2$
    C. $f(r)=8r^2+8r+1; x(r)=2r$
    D. $f(r)=8r^2+8r+1; x(r)=r/_2$
    E. $f(r)=2r^2+4r+1; ; x(r)=r/_2$

3. Write the parametric equations of a circle centered at origin and passing through (3,4).

    A. $x=5\cos\theta, y=5\sin\theta \quad 0<\theta\le2\pi$
    B. $x=3\cos\theta, y=4\sin\theta 0\le\theta\le2\pi$
    C. $x=5\cos\theta, y=5\sin\theta 0<\theta<2\pi$
    D. $x=3\cos\theta, y=4\sin\theta 0<\theta\le2\pi$
    E. $x=5\cos\theta, y=5\sin\theta \quad 0\le\theta\le2\pi$

4. Find the polar form of line $x=2$.
    A. $r=2\sec\theta$
    B. $r=2\sin\theta$
    C. $r=2\csc\theta$
    D. $r=2\cos\theta$
    E. $r=\cos\theta+2$

5. Write the rectangular equation of $r=2sin\theta$

    A. $x^2+y^2-4y=0$
    B. $x^2+y^2+2y=0$
    C. $x^2+y^2-2x=0$
    D. $x^2+y^2+2x=0$
    E. $x^2+y^2-2y=0$

6. Write the polar form of $xy=1$

    A. $r^2=sin\theta cos\theta$
    B. $r^2=csc2\theta$
    C. $r^2=2sec2\theta$
    D. $r^2=0.5sec2\theta$
    E. $r^2=2cos2\theta$

7. Find the parametric equation of the ellipse $x^2+25y^2=100$

    A. $x=100\ cos\theta;\ \ y=4\ sin\theta$ $0\leq\theta\leq2\pi$
    B. $x=10\ cos\theta;\ \ y=2\ sin\theta$ $0\leq\theta\leq2\pi$
    C. $x=cos\theta;\ \ y=5\ sin\theta$ $0\leq\theta\leq2\pi$
    D. $x=10\ cos\theta;\ \ y=5\ sin\theta$ $0\leq\theta\leq2\pi$
    E. $x=100\ cos\theta;\ \ y=25sin\theta$ $0\leq\theta\leq2\pi$

Mini-Assessment Answers and Explanations

1. Answer B

Explanation:
A parameter is a variable in which other variables are experienced in terms of. In the above equations, we see that x and y are expressed in terms of t. Thus, t is the parameter.

2. Answer C

Explanation:
From $\frac{x}{r} = 2$, we have $= x(r) = 2r$
Substituting for x in the function $f(x)$, we get
$f(r) = 2(2r)^2 + 4(2r) + 1 = 8r^2 + 8r + 1$
the answer is $f(r) = 8r^2 + 8r + 1$; $x(r) = 2r$

3. Answer E

Explanation:
The radius of the circle is the distance from the origin to the circle. That is

$$r = \sqrt{(3^2 + 4^2)} = \sqrt{25} = 5$$

Since the circle is complete, the angle ranges from 0 to 2π, that is $0 \leq \theta \leq 2\pi$

4. Answer A

Explanation:
Using polar coordinates, $x = r\cos\theta$, hence we have $r\cos\theta = 2$
Making r the subject of the formula, we get $r = \frac{2}{\cos\theta} = 2\sec\theta$

5. Answer E

Explanation
We use the polar equivalents
$r^2 = x^2 + y^2$ and $y = r\sin\theta$ hence $\sin\theta = \frac{y}{r}$
Substituting into the above equation, we get
$$r = \frac{2y}{r}$$
$$Or \quad r^2 = 2y$$
Substituting for $r2$, we get
$$x^2 + y^2 = 2y$$
$$Or \; x^2 + y^2 - 2y = 0.$$

6. Answer C

Explanation:
We have $x = r\cos\theta$ and $y = r\sin\theta$
Upon substitution, we get
$r2\cos\theta\sin\theta = 1$
But $sin2\theta = 2\cos\theta\ \sin\theta$ hence $\frac{sin2\theta}{2} = \cos\theta\ \sin\theta$
Upon substitution, we get

$$\frac{r^2 sin2\theta}{2} = 1; \quad r^2\ sin2\theta = 2$$

$$r^2 = \frac{2}{sin2\theta} = 2sec2\theta$$

Thus $r^2 = 2sec2\theta$

7. Answer B

Explanation:
We write the equation in standard form. Dividing through by 100 gives us

$$\frac{x^2}{100} + \frac{25}{100}y^2 = 1 \quad or \quad \frac{x^2}{10^2} + \frac{y^2}{2^2} = 1$$

Comparing with $\quad \frac{x^2}{a^2} + \frac{y^2}{b^2} = 1,\ $ we have $a = 10$ and $b = 2$

The parametric equations are given by $x = acos\theta; \quad y = b\ sin\theta$
Hence $x = 10\ cos\theta; \quad y = 2\ sin\theta \quad\quad 0 \le \theta \le 2\pi$

Lesson Reflection:

In this lesson, we learned about parametric and polar equations. As we close our lesson today, please reflect on the following questions:

- How would you describe parametric and polar equations?

- What was your biggest take-a-way about modeling and solving parametric and polar equations?

Lesson 3
Vectors

Lesson Description:

This lesson is designed to help students understand how vectors are represented. Also, students will analyze the properties of vectors and explore the basic vector arithmetic.

Please be sure to utilize the questions to help spark student engagement and cover the vocabulary that is associated with this specific tutoring session. For your own knowledge, sample responses have been provided to guide you as well.

Learning Objectives:

In today's lesson, the learner will describe the properties of vectors and basic vector arithmetic with 75% accuracy in 3 out of 4 trials.

Introduction

When we move in different directions from a starting to the final position, a straight line joinsour two points. However, we have a similar but slightly different situation. We move from the very starting point, make all possible turns then go to the very final point.In this two cases, the destination and starting points are the same but the directions are not. In the first case, the direction is one and specific. In the second case, the direction is not specific, in fact the distance covered is longer. The first situation is said to be a vector while the section, a scalar. Thus, a vector is motion in a specific direction. Define in your own words, these vocabulary words. In this lesson, we will discuss more about vectors.

Questions to Spark Student Engagement

- We have already dealt with vectors, but we have not defined them definitively. For instance, when you are traveling in a specific direction.
- In your own words, how would you relate change in distance over the change in time to today's topic?
- This would be velocity which is represented by a velocity vector.

Specific Vocabulary Covered

The learner needs to know these vocabulary terms by the end of the session. As a suggestion, you can have him or her write them on flashcards or even use them as visual vocabulary words.

- **Vector**
  This is a quantity, which has both magnitude and direction.

- **Head**
  This is the ending point of a vector.

- **Tail**
  This is the starting point of a vector.

- **Scalar**
  This is a quantity, which has magnitude but has no direction.

- **Vector Components**
  These are sub vectors that together make up the real vector

What are Vectors?

Vectors are quantities with direction and magnitude.

For example, this may include velocity and force.

Head of Vector

Tail of Vector

Vector Addition

- Given that *a* and *b* are vectors, then we have *b* at the head of *a*. Please note that $b + a = a + b$

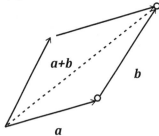

a+b

b

a

Scalar Multiplication

- A scalar, typically represented by *c*, is a real number.

- For the following depiction, the difference among the vectors is defined as *a + (-b) = a - b*

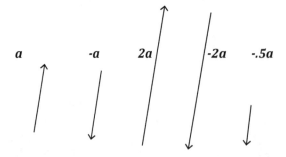

a   *-a*   *2a*   *-2a*   *-.5a*

Vector Arithmetic

Let's say we have two vectors: $a = <a_1, a_2>$ and $b = <b_1, b_2>$, and a scalar c, then we know:

$ca = <ca_1, ca_2>$

$a + b = <a_1 + b_1, a_2 + b>$

$a - b = <a_1 — b_1, a_2 — b>$

Vector Properties

If a, b, and c are vectors and d and e are scalars, then we know:

$a + b = b + a$

$a + (b + c) = (a + b) + c$

$a + 0 = a$

$a + (—a) = 0$

$d(a + b) = da + db$

$(d + e)a = da + ea$

$(de)a = d(ea)$

$1 a = a$

Guided Instruction

Let's say you have a bucket full of cement weighing **500kg** suspended by two ropes from a beam. Find the tension on the ropes.

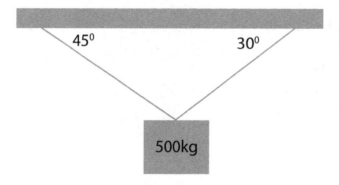

We have to begin by finding the vertical components.

$$acos(45) = bcos(30)$$

$$asin(45) + bsin(30)$$

$$a\frac{1}{\sqrt{2}} - b\frac{\sqrt{3}}{2} = 0$$

$$a\frac{1}{\sqrt{2}} + b\frac{1}{2} = 500$$

$$\text{left} = \frac{500}{\frac{\sqrt{3}}{2} + \left(\frac{1}{2}\right)} = 366kg$$

$$\text{right} = \sqrt{2}\frac{500}{\frac{\sqrt{3}}{2} + \left(\frac{1}{2}\right)} * \frac{\sqrt{3}}{2} = 448.3kg$$

Video Suggestions

While there are many videos available to help you, these are only to be used as a starting point to help you with any supplemental videos in which you may use. Please conduct a search on either YouTube or Teacher Tube to find appropriate videos for this lesson.

Below are some suggested title searches:

■ Vector Arithmetic

■ Vector Properties

Independent Instruction: Working on Your Own

Now, it's the learner's turn. This is the learner's chance to demonstrate that he or she can complete the skill on his or her own. Depending on how much time in which you have left in the session, you may want to use only one or two of these questions and focus on the mini-assessment questions that are aligned with this particular lesson. Of course, it will depend on both the session time and the learner's progress.

Let's say we have two vectors:

$a =< 3,-2,1 >$ and $b =< 2,0,-2 >$

1. Find $a+ b$.
$< 3 + 2,-2 + 0,1 + (-2) >=< 5,-2,-1 >$

2. Find $a - b$.
$< 3 - 2,-2 — 0,1 + 2 > =< 1,-2,3>$

3. Find $2a + 2b$.
$2a =< 6, —4, 2 >$ and $2b =< 4,0, —4 >$
$2a — 2b =< 2,-4,6 >$

Mini-Assessment

At the end of your tutoring session, these are the questions that the learner will need to complete. Please make sure that you keep in mind only the topics in which you have covered in the lesson. As a suggestion, if time does not permit, you can have the learner complete the rest of the mini-assessment at the next tutoring session.

1. Say you ride your bike 10 miles north, then 10 miles east, what is your total displacement?

      A. 100m   B. 20m   C. 10m   D. none of these.

2. Perhaps you are looking for your lost dog. You know he traveled 3km south and then abruptly turned and traveled 4km west. What is his total displacement?

      A. 5km   B. 25km   C. 7km   D. none of these.

3. Say that you are pulling a wagon in the magical physics vacuum where there is no friction or other outside forces. You push a box forward 5 meters, then to the left 2 meters. There is a force of 2 Newtons. What is the work done if work is equal to force times displacement?

      A. 5.33 J   B. 9 J   C. 10.77J   D. None of these.

4. What is the work done if the displacement is 16m and the force is 4N?

5. What is the displacement of an object if you move it 14 miles N, then 10 miles W?

6. What is the velocity if the displacement is 14m and the total time is 10s?

7. What is the velocity if the displacement is 2m and the time is 1s?

Mini-Assessment Answers and Explanations

1. Say you ride your bike 10 miles north, then 10 miles east, what is your total displacement?

A. 100m   B. 20m   C. 10m   D. none of these.

Answer E

Explanation:
We can sketch the movement as follows.

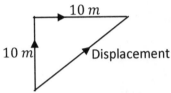

The distance between the starting point and the final point is the total displacement.

Use can find this displacement using the Pythagorean Theorem as follows. The total displacement is $\sqrt{10^2 + 10^2} = \sqrt{200} = 14.14$ miles

2. Perhaps you are looking for your lost dog. You know he traveled 3km south and then abruptly turned and traveled 4km west. What is his total displacement?

a. 5km   b. 25km   c. 7km   d. none of these.

Answer: A, 5km

Explanation:

We can sketch the dog's movement as follows.

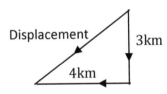

The distance between the starting point and the final point is the total displacement.

Use can find this displacement using the Pythagorean Theorem as follows. The total displacement is $\sqrt{3^2 + 4^2}$ = 5km

3. Say that you are pulling a wagon in the magical physics vacuum where there is no friction or other outside forces. You push a box forward 5 meters, then to the left 2 meters. There is a force of 2 Newtons. What is the work done if work is equal to force times displacement?

a. 5.33 J   b. 9 J   c. 10.77J   d. None of these.

Answer: C, 10.77J

Explanation:

We first find the displacement as follows

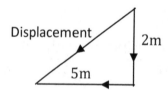

The total displacement is $\sqrt{(5^2+2^2)} = 5.385$ m

Work done = displacement × force

Work done = 5.385×2 = 10.77 J

4. What is the work done if displacement is 16m and the force is 4N?

Work done = displacement × force
Work done =16×4=64 J

5. What is the displacement of an object if you move it 14 miles N, then 10 miles west?

This movement can be sketched as follows.

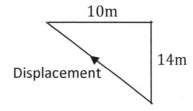

10m

14m

Displacement

The distance between the starting point and the final point is the total displacement.

Use can find this displacement using the Pythagorean Theorem as follows.
Displacement $\sqrt{14^2 + 10^2} = 17.20$ miles

6. What is velocity if the displacement is 14m and the total time is 10s?

$$\text{Velocity} = \frac{\text{displacement}}{\text{time}} = \frac{14}{10} = 1.4\text{m/s}$$

7. What is velocity if the displacement is 2m and the time is 1s?

$$\text{Velocity} = \frac{\text{displacement}}{\text{time}} = \frac{2}{1} = 2\text{m/s}$$

Lesson Reflection:

In this lesson, we learned about vectors. As we close our lesson today, please reflect on the following questions:

- How would you describe the properties of vectors?

- What was your biggest take-a-way about vectors?

Lesson 4
Rational Functions

Lesson Description:
This lesson is designed to help students describe the characteristics of rational functions and graph rational functions.

Please be sure to utilize the questions to help spark student engagement and cover the vocabulary that is associated with this specific tutoring session. For your own knowledge, sample responses have been provided to guide you as well.

Learning Objective(s)
In today's lesson, the learner will describe the characteristics of rational functions and graph rational functions with 75% accuracy in 3 out of 4 trials.

Introduction

We understand the speed the cost function is a liner function and the number of items to be considered is a simple linear function. When the later divides the former, we get a fractional like function called a rational function. Rational functions is a function can be written $f(x) = \frac{g(x)}{P(x)}$ where $g(x)$ and $p(x)$ are polynomials with $p(x)$ not being zero. In this lesson, we are interested in knowing the features of this function and attempt to sketch it.

Questions to Spark Student Engagement

- Have you seen a graph of a rational function?

- Have you ever wondered how will you be able to come up with the graph of a rational function without the use of technology?

Real-World Connections

There are many real-world connections for this particular lesson. Please make sure that you communicate this clearly to your learner prior to beginning the tutoring session.

- Knowing what rational functions are and familiarity of its characteristics and features will expand the learner's knowledge about functions
- Understanding the different features will enable learners to graph the rational function with utmost accuracy and see how the function behaves around its domain
- Learners will know that not all rational function graphs look the same

Specific Vocabulary Covered

The learner needs to know these vocabulary terms by the end of the session. As a suggestion, you can have him or her write them on flashcards or even use them as visual vocabulary words.

- **Rational Function**
  This is a function that can be represented as a fraction of polynomial functions

- **Horizontal asymptote**
  The horizontal asymptote of a function is a line $y = L$ if $lim_{x \to \infty} f(x) = L$

- **Vertical asymptote**
  The horizontal asymptote of a function is a line $x = P$ if $lim_{x \to P} f(x) = \infty$

- **Intercepts**
  Point at which a curve or a line intersects an axis.

What is a Rational Function?

Let $f(x) = \frac{2x+1}{x-1}$

- *What can you say about this function? How is this different from other functions that you know so far?*

- *What do you notice about the numerator and denominator?*

A rational function is a function in fraction form, where both numerator and denominator are polynomials.
The general form of a rational function is $f(x) = \frac{m(x)}{n(x)}$.

m(x) & *n(x)* are both polynomials.

What are the characteristics of Rational Functions?

Since a rational function is a fraction, it follows the limitation of fractions, ie. the denominator cannot be equal to *0 (n(x) # 0)*

The root/s of the denominator (ie. the x-values that will make the denominator equal to 0) are also the **vertical asymptotes** of the function

A **vertical asymptote** is a vertical line with equation $x = a$, in which the function increases or decreases on one or both sides of the line as x approaches a.

A rational function also has a **horizontal asymptote**, which is a horizontal line with equation $y = b$, such that as x values of the graph increase or decrease, the graph approaches the line $y = b$

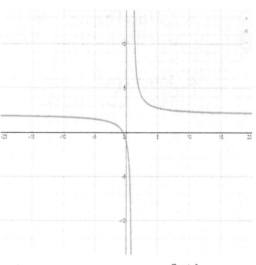

Looking at the graph on the right, it has a vertical asymptote at $x = 1$ and a horizontal asymptote at $y = 2$

Graph of $f(x) = \frac{2x+1}{x-1}$

Graphing Rational Functions

Graphing rational functions without the use of technology (ie. by hand) can be a little tricky. In order to sketch the graph as accurate as possible, the following components need to be identified:

1. Axis intercepts

These are the x-intercept/s and y-intercept of the graph

To find these:

x-intercepts set $y = 0$, then find x
y-intercept set $x = 0$, then find y

Let's take the example earlier: $f(x) = \dfrac{2x+1}{x-1}$

To get the **x-intercepts** set $y = 0$:

$$0 = \dfrac{2x+1}{x-1} \rightarrow \quad \text{a fraction is equal to 0}$$

when the numerator is 0
$0 = 2x + 1$
$-1 = 2x$
$\dfrac{-1}{2} = x$ $\quad \therefore$ point $(\dfrac{-1}{2}, 0)$

To get the **y-intercepts** set $x = 0$:
$y = \dfrac{2(0)+1}{0-1}$
$y = \dfrac{0+1}{-1}$
$y = \dfrac{1}{-1}$
$y = -1,$ $\quad \therefore$ point $(0, -1)$

2. Vertical asymptote/s
To get the vertical asymptote, the denominator function is set to 0, then solve for x

$f(x) = \dfrac{2x+1}{x-1}$, the denominator is $x - 1$
$x - 1 = 0$
$x = 1$

So far, we have the following sketches:

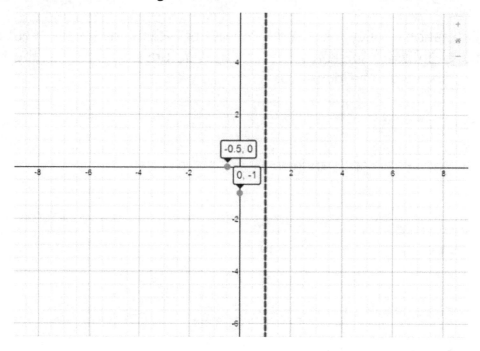

3. Horizontal asymptote

To get the horizontal asymptote, the leading exponents of numerator and denominator will be compared

The general form of a rational function is as follows:

$$f(x) = \frac{ax^m + \ldots}{bx^n + \ldots},$$

where **m** and **n** are the leading exponents of the numerator and denominator, respectively

The horizontal asymptote (HA) is identified based on the following conditions:

If **m** = **n**, HA = $\frac{a}{b}$

If **m** < **n**, HA = 0

If **m** > **n**, there's no horizontal asymptote

For the given function, $f(x) = \frac{2x+1}{x-1}$, the leading exponents are x-i both 1, therefore, $HA = \frac{2}{1} = 2$. Given this, we have the following sketch:

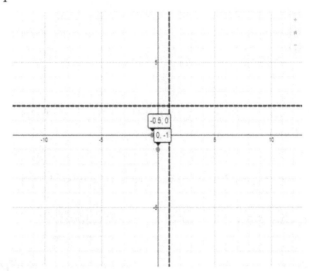

As can be seen, the asymptotes divided the Cartesian plane into four quadrants. The graph of the rational function will be on opposite, alternating quadrants. Since the axis intercepts are on the lower left quadrant, it means that the other part of the graph is on the upper right quadrant. The graph approaches the asymptotes but do not touch them.

Hence, the graph of function, $f(x) = \frac{2x+1}{x-1}$, looks like:

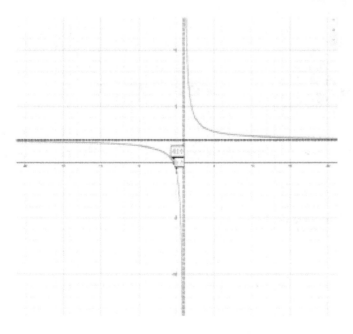

Notice how the graph approaches the asymptotes as x and y increases and decreases Since only the lower left quadrant has points plotted, the upper right part of the graph may be estimately drawn; but if more accuracy is desired, random points within the domain/range can be plotted also.

Now, let's have another example: $f(x) = \dfrac{x + 2}{2x^2 - 6x + 4}$

How different is this from the previous example?

Take note that the denominator for this rational function is a quadratic function.

Knowing that the denominator is in quadratic form, what do you think is the effect on the intercepts and asymptotes of the graph?

Is the graph similar to how previous example looks like? Or is it totally different?

Let's start identifying the intercepts and asymptotes:

- *x-intercept:*

$$f(x) = \frac{x+2}{2x^2-6x+4}$$

$$0 = \frac{x+2}{2x^2-6x+4}$$

$$0 = x + 2$$

$$x = -2, \therefore point(-2, 0)$$

- *vertical asymptote/s:*

$$2x^2 - 6x + 4 = 0$$

factoring the above quadratic function:

$$(2x - 2)(x - 2) = 0$$

$$x = 1 \; or \; x = 2$$

- *y-intercept:*

$$f(x) = \frac{x+2}{2x^2-6x+4}$$

$$y = \frac{0+2}{2(0)^2-6(0)+4}$$

$$y = \frac{2}{4} = \frac{1}{2}, \therefore point\left(0, \frac{1}{2}\right)$$

- *horizontal asymptote:*

$$f(x) = \frac{x+2}{2x^2-6x+4}$$

leading coefficients:

$$m = 1; n = 2$$

$$m < n, \therefore HA = 0$$

Now that we have the basic components of the rational function, let's see if we can already graph the function

$$f(x) = \frac{x + 2}{2x^2 - 6x + 4}$$

So far, we have the follioing sketches:

Since there are two vertical asymptotes, this means the plane is divided into 6 boxes

Both intercepts are located on the upper left box, but that does not tell us which boxes the other parts will be

Also, note that the horizontal asymptote is at $y = 0$, but there is an x-intercept which crosses the x-axis $(y = 0)$

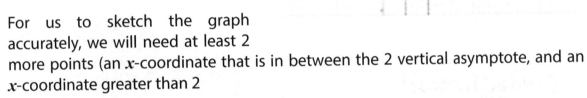

For us to sketch the graph accurately, we will need at least 2 more points (an *x*-coordinate that is in between the 2 vertical asymptote, and an *x*-coordinate greater than 2

For a more accurate sketch of the graph, let's get two more points where the graph passes though:

Point 1: let $x = \frac{3}{2}$ *(this point is in between the two vertical asymptotes)*

$$y = \frac{^3/_2 + 2}{2(^3/_2)^2 - 6(^3/_2) + 4}$$

$$y = \frac{^7/_2}{^9/_2 - 9 + 4} = \frac{^7/_2}{-^1/_2} = \frac{7}{2} \times -\frac{2}{1}$$

$$y = -7, \therefore point\left(\frac{3}{2}, -7\right)$$

Point 2: let $x = 4$ *(greater than the rightmost vertical asymptote)*

$$y = \frac{4+2}{2(4)^2 - 6(4) + 4} = \frac{6}{32 - 24 + 4}$$

$$y = \frac{6}{12} = \frac{1}{2}, \therefore point\left(4, \frac{1}{2}\right)$$

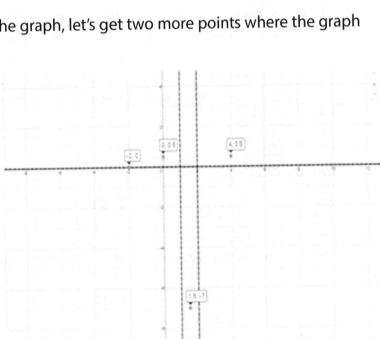

Given these points and knowing the asymptotic behavior of the graph, a more accurate graph can now be sketched representing the function:

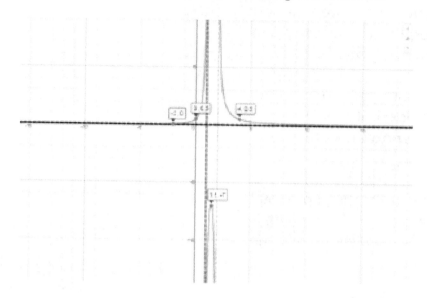

Question 1:

Consider the function $f(x) = \dfrac{4 - 2x}{x - 1}$

Identify the asymptotes, intercepts and sketch the graph.

- Just by looking at the function, are you able to identify the vertical asymptote of the function?
- What are the leading exponents of the numerator and denominator? What then is the horizontal asymptote?
- Which quadrant do the intercepts lie?
- How does this function behave as x increases or decreases?

Question 2:

Completely sketch the function $f(x) = \dfrac{2x}{x^2 - 4}$ showing all the important features.

- Just by looking at the function, are you able to identify the horizontal asymptote of the function?
- How many vertical asymptotes does this graph have?
- How many x-intercepts does this graph have?
- How many segments/parts does the graph of this function have?

Video Suggestions

While there are many videos available to help you, these are only to be used as a starting point to help you with any supplemental videos in which you may use. Please conduct a search on either YouTube or Teacher Tube to find appropriate videos for this lesson.

Below are some suggested title searches:

- Rational Functions

Independent Instruction: Working on Your Own

Now, it's the learner's turn. This is the learner's chance to demonstrate that he or she can complete the skill on his or her own. Depending on how much time in which you have left in the session, you may want to use only one or two of these questions and focus on the mini-assessment questions that are aligned with this particular lesson. Of course, it will depend on both the session time and the learner's progress.

Question 1:

Completely sketch the function $f(x) = \dfrac{3x-2}{x+3}$ showing all the important features.

Explanation:

- Just by looking at the function, we can identify the vertical asymptote to be

- Since the leading exponents are equal for both numerator and denominator (both are 1), HA is

- Plugging in 0 for x, then for y, we get the intercepts as: $\left(\dfrac{2}{3}, 0\right)$ and $\left(0, -\dfrac{2}{3}\right)$

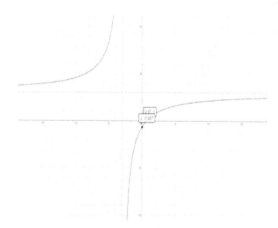

Question 2:

Completely sketch the function $f(x) = \dfrac{x-2}{x^2+x-2}$ showing all the important features.

Explanation:

- The leading exponent of the numerator is 1, while the denominator is 2; from the HA rules, this means that HA is y = 0

- Since the denominator is a quadratic function and has 2 roots: x = -2 and x=1, this means that the function also has 2 VAs at those points

- Plugging in 0 for x, then for y, we get the intercepts as: (2,0) and (0,1)

Question 3:

Completely sketch the function $f(x) = -2 + \dfrac{1}{4-x}$ showing all the important features.

Explanation:

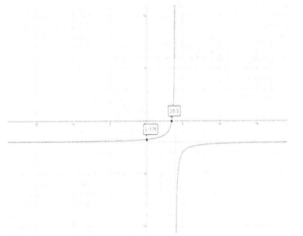

- This function is a bit different from the others as it comes with a whole number, which should just be treated as a vertical shift of the graph of the rational expression

- The features of the rational expression $\dfrac{1}{4-x}$ are obtained the same way, except for HA:

VA: $x = 4$ **HA:** $y = -2$ because of the vertical shift down of -2

Axis intercepts: (3.5 , 0) and (0, -1.75)

Mini-Assessment

At the end of your tutoring session, these are the questions that the learner will need to complete. Please make sure that you keep in mind only the topics in which you have covered in the lesson. As a suggestion, if time does not permit, you can have the learner complete the rest of the mini-assessment at the next tutoring session.

- Develop 7 word problems that incorporates the skills in today's lesson.
- Please have 3 with Multiple Choice (MC) and 4 without Multiple Choice.

For questions 1-2, consider the function:

$$\frac{9 - x^2}{(x - 1)(x + 2)}$$

1. What are the equations of the vertical asymptotes of this function?
    a. X = 1   and   x = 2
    b. X = -1   and   x = -2
    c. X = 1   and   x = -2

2. What is the equation of the horizontal asymptote of this function?
    a. Y = 0
    b. Y = 1
    c. Y = -1

Sketch the graph of the rational functions in questions 5-7:

5. $f(x) = \dfrac{3x - 3}{(x + 2)^2}$

6. $f(x) = \dfrac{3x}{2x - 5}$

7. $f(x) = 3 - \dfrac{4}{2x - 5}$

For questions 1-2, consider the function:

$$\frac{9 - x^2}{(x - 1)(x + 2)}$$

1. What are the equations of the vertical asymptotes of this function?

    a. $X = 1$ and $x = 2$

    b. $X = -1$ and $x = -2$

    c. $X = 1$ and $x = -2$

Answer C

Explanation:

At vertical asymptote, we have $(x-1)(x+2) = 0$; $x - 1 = 0$ and $x + 2 = 0$
Thus, $x=1$ and $x=2$

2. What is the equation of the horizontal asymptote of this function?

    a. $Y = 0$

    b. $Y = 1$

    c. $Y = -1$

Answer B

Explanation:

On expanding the denominator, we get $x^2 + x-2$

The numerator and the denominator have the same degree hence the horizontal asymptote is the ratio of the leading coefficient. That is $y=1$

Sketching the graphs of rational functions

5. $f(x) = \dfrac{3x-3}{(x+2)^2}$

The intercepts
At y-inercept, $x=0$, thus we substitute 0 for x.

$f(0) = -\dfrac{3}{2^2}$, this gives us $y = -\dfrac{3}{4}$ $= -0.75$ when $x = 0$.

x- intercept
At x-intercept, $y=0$. $3x-3 = 0$, $x=1$

The x-intercept is 1.
The intercepts are $(1, 0)$ and $(0, -0.75)$

Vertical asymptote
We equate the denominator to zero and solve for x

Solving $(x+2)^2 = 0$ we get $x=-2$.

The line $x =-2$ is therefore a vertical asymptote.
Since the degree of numerator is less than that of the denominator, the line $y=0$ is horizontal asymptote.

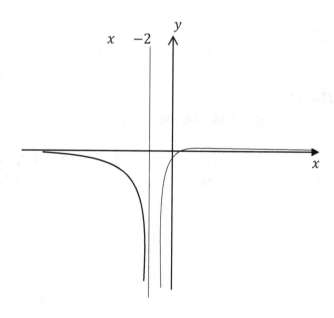

Sketching the graphs of rational functions

6. $f(x) = \dfrac{3x}{2x-5}$

x-intercept
At this point $y=0$, hence $3x=0$, $x=0$.

y-intercept
At this point, $x=0$, hence $y=0$

Vertical asymptote
$2x - 5 = 0$; $x = \dfrac{5}{2} = 2.5$

Horizontal asymptote
The numerator and the denominator have equal degree, hence the horizontal asymptote is $y = \dfrac{3}{2} = 1.5$

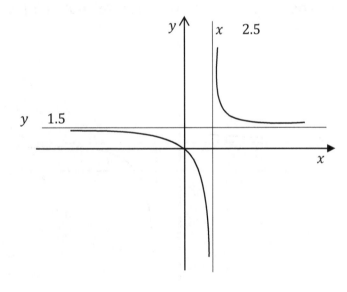

Sketching the graphs of rational functions

7. $f(x) = 3 - \dfrac{4}{2x - 5}$

x-intercept
At this point $y=0$, hence $3 - \dfrac{4}{2x - 5} = 0;$ $\dfrac{4}{2x - 5} = 3$

y-intercept
At this point, $x=0$, thus, $y = 3 - \dfrac{3}{2} = 2.2$

Vertical asymptote
$2x - 5 = 0;$ $x = 2.5$

Horizontal asymptote

$$f(x) = 3 - \dfrac{4}{2x - 5}$$

The asymptote here is $y=3$, which is horizontal, not oblique

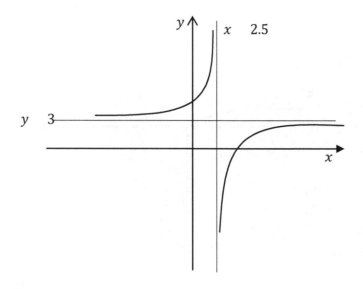

Lesson Reflection

In this lesson, we learned about rational functions. As we close our lesson today, please reflect on the following questions:

- How would you describe rational functions?

- What was your biggest take-a-way about graphing rational functions?

Lesson 5
Square Root Functions

Learning Objectives:

This lesson is designed to help students identify square root functions and the domain and range of square root functions. Additionally, learners will graph transformations of square root functions.

Please be sure to utilize the questions to help spark student engagement and cover the vocabulary that is associated with this specific tutoring session. For your own knowledge, sample responses have been provided to guide you as well.

Learning Objective:

In today's lesson, the learner will identify and graph square root functions with 75% accuracy in 3 out of 4 trials.

Introduction

When we have a square piece of in mind and whose area is known, one would be thinking of how to fence it to secure it from trespassers. Fencing implies using the determining its length then knowing the amount of material that would be required. This is a typical case of square root of functions. Determining the side of the square piece of land is a simple as determining the square root of the area. In this lesson, we are going to learn how to determine the square root of numbers.

Questions to Spark Student Engagement

- Have How are square root functions different/similar to square root of numbers?
- Do you think you will be able to graph a square root function manually (ie without the use of a graphing calculator)?
- How are square root functions different/similar to square root of numbers?
  - *The square root of numbers are just constants or pure numbers inside the square root symbol; while a square root function has variable x and a polynomial function inside the square root.*

- Do you think you will be able to graph a square root function manually (ie without the use of a graphing calculator)?
  - *I'm not sure. I think it's difficult.*

Real-World Connections

There are many real-world connections for this particular lesson. Please make sure that you communicate this clearly to your learner prior to beginning the tutoring session.

- Knowing what a square root function is will expand the learner's knowledge about different types of functions
- Understanding the domain and range of a square root function will help the learners draw its graph without the use of technology
- Learners will know the different transformations possible for a square root function and will be able to graph them

Specific Vocabulary Covered

The learner needs to know these vocabulary terms by the end of the session. As a suggestion, you can have him or her write them on flashcards or even use them as visual vocabulary words.

- *Square Root Function*
  this is a type of function that is inside the square root symbol

- *Domain and Range*
  domain is the set of all possible x-values of the function; range is the set of all possible y-values of the function

- *Horizontal and Vertical Stretch*
  horizontal stretch is the constant multiplier for x, while vertical stretch is the constant multiplier for y; these changes the graph by compressing or expanding the graph

- *Horizontal and Vertical Shift*
  horizontal and vertical shifts are the movement of the graph left or right and up or down, respectively

- *Reflection*
  this is the translation over the opposite side of the x- or y-axis

What is a Square Root Function?

Let $f(x) = \sqrt{x}$

- What can you say about the square root function?

- What do you think the graph of this function looks like?

- Are there any restrictions to this function?

A square root function contains a radical expression (ie a square root) with the variable x in the radicand.

The square root function is the inverse of a quadratic function.

$f(x) = \sqrt{x}$ is the inverse of $f(x) = x^2$

Graph of a Square Root Function

To better understand the characteristics of a square root function, particularly its domain and range, let's look at the graph of the parent function $f(x) = \sqrt{x}$

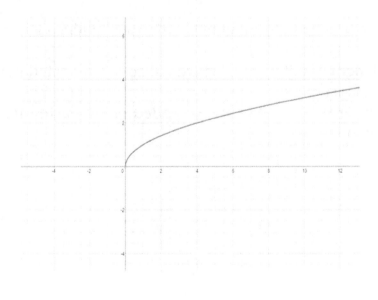

- The graph of $f(x) = \sqrt{x}$ is seen on the left

- It is noticeable that the graph is only on the positive values of x and y

- Note that even though square roots of numbers will result in both positive and negative values, the negative y-values are not included as it will render the expression a non-function

It was mentioned previously that a square root function is the inverse of a quadratic function. Let's validate this by looking at the graph of the two parent functions.

- The function $f(x) = \sqrt{x}$ is a reflection of the function $f(x) = x^2$ over the $y = x$ line

- Given the graph of the square root function, the domain and range can be defined as follows:

  Domain: $\{x \mid x \geq 0\}$
  Range: $\{y \mid y \geq 0\}$

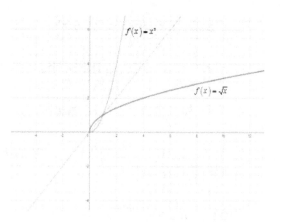

Transformations

Now, let's look at the different transformations possible with square root functions. The standard form of a square root function showing the transformations is as follows:

$$f(x) = a\sqrt{x - h} + k$$

where

a = **vertical stretch** (ie. y-values of the parent function are multiplied by the factor a)

h = **horizontal shift**, right or left (ie. *h* units are added or subtracted from x-values of the parent function)

k = **vertical shift**, up or down (ie. *k* units are added or subtracted from y-values of the parent function)

Graphing Square Root Functions

To show the transformation graphically, let's take the example of the function,

$$f(x) = \sqrt{x-3} + 4$$

The function above has two transformations from the parent function:

- a **horizontal shift** of _3 units to the right_
- a **vertical shift of** _4 units up_

The graph of the function is seen on the right

Notice that the initial point (0,0) has been transformed into (3,4); therefore, the domain and range are also shifted

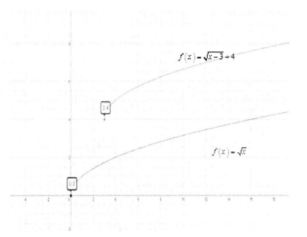

Domain: {x | x ≥ 3}
Range: {y | y ≤ 3}

Let's look at another example:

$$f(x) = 2\sqrt{x+1} - 3$$

- *How many transformations happened in this function? Can you name all of them?*

- *What happens to the y-values of the function? How about to the x-values?*

There are 3 transformations on this function:

- **vertical stretch** of _2 units_
- **horizontal shift** of _1 unit to the left_
- **vertical shift** of _3 units down_

Sample student responses:

- How many transformations happened in this function? Can you name all of them?

    There are three transformations, the vertical stretch of 2, the horizontal shift of 1 unit to the left and vertical shift of 3 units down.

- What happens to the y-values of the function? How about to the x-values?

    The y-values are multiplied by 2 and subtracted 3 points, while x values are subtracted 1 point.

The graph of the function $f(x) = 2\sqrt{x+1} - 3$ is shown bellow:

- The parent function is vertically stretched by 2 units making the new graph wider

- The initial point also moved from (0,0) to (-1,-3)

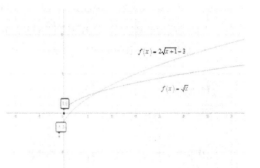

Domain: $\{x \mid x \le 1\}$ **Range:** $\{y \mid y \ge -3\}$

Let's have another example:
$$f(x) = -\sqrt{2-x}$$

Is this a valid function? If so, what transformations happened to this function?

- The above function is also valid as it is a transformation from the parent function $f(x) = \sqrt{x}$

- Let's first rewrite the function reflecting the standard form:
$$f(x) = \sqrt{-(x-2)}$$

The transformations are as followings:

    the negative <u>outside</u> the radical is a **reflection** <u>about the x-axis</u>

    the negative <u>inside</u> the radical is a **reflection** <u>about the y-axis</u>

    horizontal shift of *2 units to the right*

The graph of the function $f(x) = \sqrt{-(x-2)}$ is shown below:

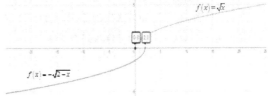

- The parent function is reflected about the *x*-axis, then about the *y*-axis

- The initial point also moved from (0,0) to (2,0)

Domain: $\{x \mid x \le 2\}$
Range: $\{y \mid y \le 2\}$

Question 1:

Consider the function $f(x) = 3\sqrt{x+2} + 4$

Identify the transformations from the parent function $f(x) = \sqrt{x}$ and sketch the graph of the given function.

- What are the different stretch and shifts that happened in this function?
- How does the graph of this function compare with the parent function?
- What is the new coordinate of the initial point (0,0)?

Question 1 Sample Student Response:

- What are the different stretch and shifts that happened in this function?
  - This function has vertical stretch of 3 units, horizontal shift of -2 units and vertical shift of 4 units

- How does the graph of this function compare with the parent function?
  - The graph of this function moved left and up, and it is vertically stretched so it is wider (vertically) than the parent function.

- What is the new coordinate of the initial point (0,0)?
  - The new coordinate is (-2,4).

The graph of the function is shown below:

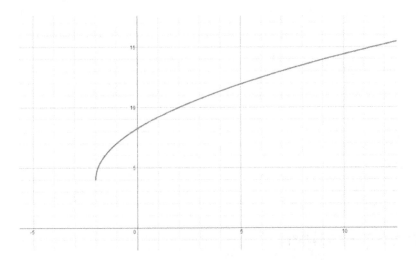

Question 2:

Completely sketch the function $f(x) = -\sqrt{x} - 5$ and identify its domain and range.

- What are the different transformations that happened here?
- What is the minimum x-value of the function?
- What is the minimum y-value of the function?

Question 2 Sample Student Response:

- **What are the different transformations that happened here?**
  - This function is reflected over the x-axis and moved down 5 units

- **What is the minimum x-value of the function?**
  - The minimum x-value of the function is x=0

- **What is the minimum y-value of the function?**
  - This graph has no minimum y-value, but has a maximum value of y=-5

The graph is shown on the right.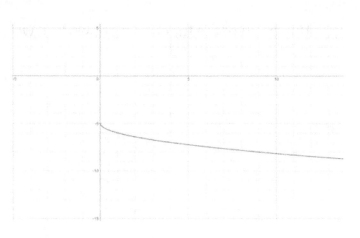

The domain is $\{x \mid x \geq 0\}$

The range is $\{y \mid y \leq -5\}$

While there are many videos available to help you, these are only to be used as a starting point to help you with any supplemental videos in which you may use. Please conduct a search on either YouTube or Teacher Tube to find appropriate videos for this lesson.

Below are some suggested title searches:

- Graphing Square Root Functions

- Radical Functions

Independent Instruction: Working on Your Own

Now, it's the learner's turn. This is the learner's chance to demonstrate that he or she can complete the skill on his or her own. Depending on how much time in which you have left in the session, you may want to use only one or two of these questions and focus on the mini-assessment questions that are aligned with this particular lesson. Of course, it will depend on both the session time and the learner's progress.

Question 1:

Completely sketch the function $f(x) = \sqrt{x+3} + 5$ and identify the domain and range.

Explanation (sample response):
There are two transformations that happened in this function:

- **horizontal shift** of <u>3 units to the left</u>
- **vertical shift** of <u>5 units up</u>

With these transformations, the initial point (0,0) will be shifted to the new coordinates (-3,5)

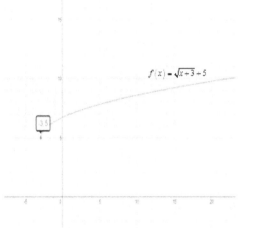

Domain: $\{x \mid x \geq -3\}$ **Range:** $\{y \mid y \geq 5\}$

Question 2:

Consider the function $f(x) = \frac{1}{2}\sqrt{x-2} - 1$

Identify the transformations and sketch the graph of the function.

Explanation (sample response):

This function transformed the parent function by a vertical stretch of $\frac{1}{2}$ and shifts 2 units to the right and 1 unit down

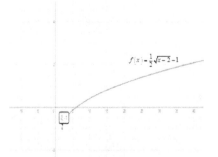

Take note of the y-axis scale, which is different from the previous graphs

Question 3:

Completely sketch the function $f(x) = -\sqrt{-4-x} + 1$ showing the transformation of (0,0) point and identifying the domain and range of the function.

Explanation (Sample Response):

Before graphing rewrite the function in standard form: $f(x) = -\sqrt{-4-x} + 1$

The rewritten function shows reflection about the x-aris, then a reflection about y-axis, a shift 4 units to the left and 1 unit up

Domain: $\{x \mid x \le -4\}$ **Range:** $\{y \mid y \le 1\}$

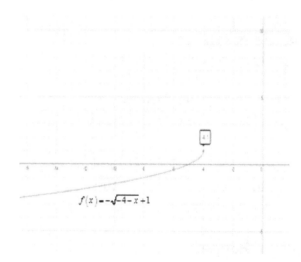

$f(x) = -\sqrt{-4-x} + 1$

Mini-Assessment

At the end of your tutoring session, these are the questions that the learner will need to complete. Please make sure that you keep in mind only the topics in which you have covered in the lesson. As a suggestion, if time does not permit, you can have the learner complete the rest of the mini-assessment at the next tutoring session.

For questions 1-2, consider the function:

$$f(x) = -\sqrt{6-x} - 5$$

1. What is the domain of this function?
     a. $\{x \mid x \leq 6\}$
     b. $\{x \mid x \leq -6\}$
     c. $\{x \mid x \geq 6\}$

2. What is the range of this function?
     a. $\{y \mid y \geq -5\}$
     b. $\{y \mid y \geq 5\}$
     c. $\{y \mid y \leq -5\}$

For questions 3-4, consider the function:

$$f(x) = \sqrt{x + \frac{3}{4} - \frac{5}{9}}$$

3. The point (0,0) on the parent function $f(x) = \sqrt{x}$ is transformed into which of the follwing coordinates on the given function above:

     a. $\left(\frac{3}{4}, -\frac{5}{9}\right)$ **b.** $\left(-\frac{3}{4}, -\frac{5}{9}\right)$ **c.** $\left(-\frac{3}{4}, \frac{5}{9}\right)$

4. Describe all transformations in this function.

Sketch the graph of the square root functions in equations 5-7

5. $f(x) = \sqrt{-x} + \dfrac{2}{3}$

6. $f(x) = -\sqrt{-x+3} + 4$

7. $f(x) = 6 - \sqrt{5+x}$

Mini-Assessment Answers and Explanations

1. A

Since x is negative, this graph is reflected over the y-axis. There is also horizontal translation of 6 units to the right. Hence, the graph now starts at x=6 and going in the negative direction. The domain, then, is {x | x ≤ 6}

2. C

This graph is also reflected over the x-axis and vertically translated 5 units down. Hence, the range is {y | y ≥ 5}

3. B

There is vertical translation of -5/9 and horizontal translation of -3/4. Hence, the point(0,0) is translated to $\left(-\dfrac{3}{4}, -\dfrac{5}{9}\right)$

4. Transformations include vertical shift of $^5/_9$ units down and horizontal shift of $^3/_4$ units left

The graphs of the functions are as follows:

5.

6.

7.

Lesson Reflection:

In today's lesson, we've learned:

- what a square root function is and its characteristics.
- how to identify the domain and range of a square root function.
- how graph a square root function manually, including different transformations applied to the parent function.

Lesson 6
Trigonometric Functions

Lesson 6 Description
This lesson is designed to help students identify the characteristics of basic trigonometric functions and graph trigonometric functions.

Please be sure to utilize the questions to help spark student engagement and cover the vocabulary that is associated with this specific tutoring session. For your own knowledge, sample responses have been provided to guide you as well.

Learning Objective(s)
In today's lesson, the learner will identify and graph trigonometric functions with 75% accuracy in 3 out of 4 trials.

Introduction

When a propeller of a helicopter moves, we can determine it rate of motion by considering the angle through which it sweeps from the beginning point to a given point. More precisely, we can describe the motion of a specific point on it in terms of the angle moved and the distance from the central point. This is a common case for designers of this propeller. To be able to do so, the rates are expressed in terms of ratios called trigonometric ratios. This along with other situations makes it vital for us to understand what the trigonometric functions are. This lessons explores this concept.

Questions to Spark Student Engagement

- How do we represent functions that are sinusoidal in nature?
- What real-world applications have sinusoidal characteristics?
- Are the trigonometric functions related to each other?

Sample student responses:
How do we represent functions that are sinusoidal in nature?
- These are represented by periodic functions
What real-world applications have sinusoidal characteristics?
- Temperature changes throughout the year; high tides and low tides

Are the trigonometric functions related to each other?
- Yes, because they can be written in terms of another function.

Real-World Connections:

There are many real-world connections for this particular lesson. Please make sure that you communicate this clearly to your learner prior to beginning the tutoring session.

- Knowing trigonometric functions will expand the learner's knowledge about different types of functions

- Understanding the characteristics of a trigonometric function will help the learners draw its graph, and its transformations, without the use of technol

- Learners will be able to apply trigonometric functions into some real-world scenarios

Specific Vocabulary Covered

The learner needs to know these vocabulary terms by the end of the session. As a suggestion, you can have him or her write them on flashcards or even use them as visual vocabulary words.

- **Trigonometric Function**
  type of function where angle is the independent variable; the dependent variables are the trigonometric ratios such as sine and cosine

- **Sine and Cosine**
  these are the 2 basic trigonometric ratios; these two corresponds to the vertical and horizontal components of a polar coordinate

- **Amplitude**
  this refers to the height (or distance of the highest point) of the function from the principal axis

- **Principal Axis**
  this is the horizontal axis of symmetry of the trigonometric function

- **Period**
  this measures one complete cycle of the trigonometric function

What are Trigonometric Functions?

Do you recall trigonometric ratios?
Do you remember how the ratios are defined?

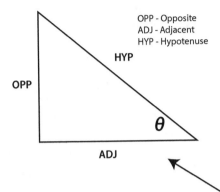

OPP - Opposite
ADJ - Adjacent
HYP - Hypotenuse

Consider the triangle on the left side. The trigonometric ratios represents the relationships of the angle, θ, with the different sides of the triangle.

The basic trigonometric ratios are sine and cosine. The are defined as follows:

$$\sin\theta = \frac{OPP}{HYP} \qquad \cos\theta = \frac{ADJ}{HYP}$$

Sample student responses:

- **Do you recall trigonometric ratios?**
  - Yes, these are sine and cosine

- **Do you remember how the ratios are defined?**
  - The pnemonic SOHCAHTOA defines the ratios sine and cosine

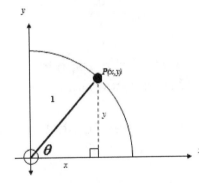

Now, let's recall the concept of the unit circle. Consider the quarter circle below: Applying the trig ratios to the figure shown, we will have the following ratios:

$$\sin\theta = \frac{OPP}{HYP} = \frac{y}{1} = y \qquad \cos\theta = \frac{ADJ}{HYP} = \frac{x}{1} = x$$

Therefore, the point *P(x, y)* can be expressed as *P(cosθ, sinθ)*.

This means that x is a function of the angle, *θ*, by way of *cosine*, and y is also a function of A by way of *sine*.

Hence, these trigonometric ratios can be expressed as **trigonometric functions** where the angle is the independent variable and the *sine* and *cosine* values are the dependent variables.

Thus, the basic trigonometric functions are expressed as follows:

$$f(x) = \sin x \qquad\qquad f(x) = \cos x$$

where x = angle, in degrees or radians

Sine Function

The table of values and points plotted on the graph are shown below:

x (degrees)	f(x)
0	0
45	$\frac{\sqrt{2}}{2}$
90	1
180	0
270	-1
360	0

Looking at the plotted points on the graph, it can be seen that the graph follows a sinusoidal pattern, like a wave. The graph of the parent function then looks like this:

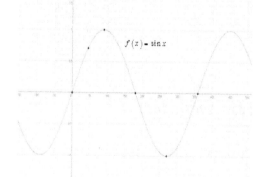

Looking at the graph on the left, what are the maximum and minimum points?

Given the cyclic behavior of the graph, what are the range of t - values where it completes one full cycle?

Sample student responses:

- Looking at the graph on the left, what are the maximum and minimum points?
    - The maximum point is 1 and minimum point is -1.

- Given the cyclic behavior of the graph, what are the range of x-values where it completes one full cycle?
    - It can be seen that the graph starts its full cycle at x=0 and ends at x=360

With the observations noted, we can enumerate the characteristics of a sinusoidal function, particularly the $f(x) = sin\ x$ function:

$f(x) = \sin x$

following a cyclic behavior and a wave-like pattern, the graph follows a cycle from $x = 0$ to $x = 360$; this characteristic is called period, which defines one full cycle on the graph

the **principal axis**, which is the horizontal line that is equidistant from the maximum and minimum points, in sine graph is $y = 0$

the **amplitude** is the height of the graph from the principal axis, and in the sine graph, it is 1

the **domain** and **range** of the sine function are
$\{x\ |x \in i\}$ and $\{y\ |\ 1 \leq y \leq -1\}$

Cosine Function

Now, let's look at the other basic trigonometric function, $f(x) = cos\ x$

The table of values and points plotted on the graph are shown below:

x (degrees)	f(x)
0	1
45	$\sqrt{2}/2$
90	0
180	-1
270	0
360	1

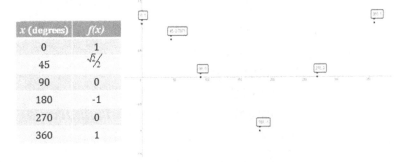

Drawing the graph of the parent function, through the plotted points, we can see the $f(x) = cos\ x$ graph as follows:

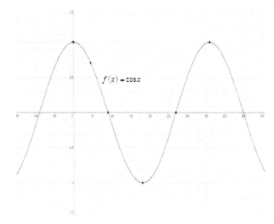

Knowing the different characteristics of trig functions, what is the period of the cosine function?

What about its principal axis and amplitude?

Can you define the domain and range of the cosine function?

Sample student responses:

- Knowing the different characteristics of trig functions, what is the period of the cosine function?
    - Period is 360

- What about its principal axis and amplitude?
    - Principal axis is at $y=0$ and amplitude is 1

- Can you define the domain and range of the cosine function?
    - Domain and range are the same as that of the sine function

Transformation

The standard form of a trigonometric function is
$$f(x) = asin\ [b(x - c)] + d$$

where a, b, c, and dare transformations of the parent function. The same standard form also applies to cosine.

The different transformations are defined as follows:
a= vertical stretch; this is also the **amplitude**
b= horizontal stretch; factor of period
c= horizontal shift (right or left)
d= vertical shift (up or down); this is also the **principal axis**

Let's look at an example. Consider the function:

f (x) = 2 sin[2xj -1

What are the transformations in this function?

There's a vertical stretch of 2 units, a horizontal stretch of 2 units and vertical shift of 1 unit down. Hence, the graph looks like:

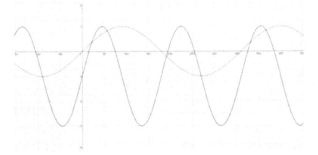

We can graph the function through the following steps:

1. Graph the parent function, *sin x*

2. Apply the horizontal stretch *sin [2x]*

3. Apply the vertical stretch *2 sin 2x*

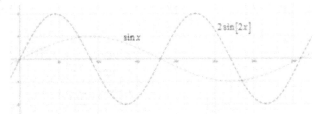

4. And lastly, move the graph 1 unit down, *2 sin [2x] - 1*

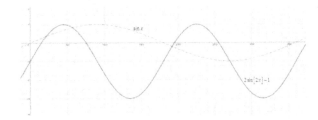

the order in which the transformations are graphed may vary. Normally, the stretches are done first before the directional shifts, but it can also be done vice versa.

Trigonometric functions can be applied in different real-world scenarios such as on the rise and fall of temperatures throughout the year, or on high tides and low tides in a day. Any situation that has a cyclic behavior in a given period may be represented through a trigonometric function.

Let's consider this scenario:

The average maximum temperatures, in °F, in City A have been recorded from January to December of last year. The table below shows these recordings:

Month	JAN	FEB	MAR	APR	MAY	JUN	JUL	AUG	SEP	OCT	NOV	DEC
Temp, °F	39	40	48	61	70	81	86	83	77	65	54	41

Given the scenario, let's find a <u>cosine function</u> that best models the given data.

So how do we approach this problem?

First let's plot the points to visualize the behavior of the data

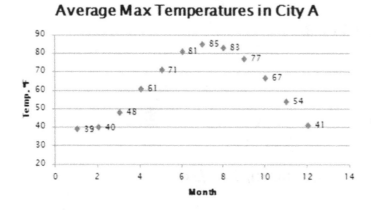

It's observable that the data follows a sinusoidal curve. Since we want to represent this data usthg a cosine function, we should compare how the parent function looks like as opposed to this.

The next step is to identify the characteristics of the parent function, cosine, and see how it relates to the given graph

We know that the function, $f(x) = cosx$, is at its maximum point when $x = 0$, which can serve as our starting reference

Looking at the graph, we see that it's at its minimum point at 39°F when $x = 1$ (January); this means that parent function is moved to the right by 1 uni4 and since the starting point for this graph is the minimum point, then we can say that the parent function is reflected over the *x-axis*

From observation, we can say that the <u>period</u> of this graph is 12; from this we can get the horizontal stretch, b, by this formula

$$b = \frac{2\pi}{period} = \frac{2\pi}{12} = \frac{\pi}{6}$$

The horizontal stretch here is expressed in radians since the dependent variable (months) is in numeral form. This can also be expressed in degrees (angles).

For the vertical stretch and vertical shift, we identify the <u>amplitude</u>, a, and the principal axis, d, using the following formulas:

$$a = \frac{\text{max} - \text{min}}{2} = \frac{85 - 39}{2} = 23 \qquad d = \frac{\text{max} + \text{min}}{2} = \frac{85 + 39}{2} = 62$$

Max and min are the maximum and minimum values from the given data points.

In summary, we have the following transformations from the parent function, cosine:

$a = 23$, and since the graph is reflected over thex-axis, the amplitude will be negative; hence, $a = -23$
$b = \pi/6$
$c = 1$ (1 unit to the right)
$d = 62$

Thus, the function that best represents the data is

$$f(x) = -23\cos\left[\frac{\pi}{6}(x-1)\right] + 62$$

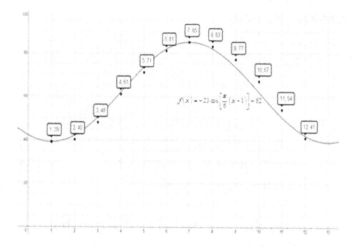

To validate, let's graph the resulting function and see how well it fits the data points.

As expected, not all data points will lie exactly on the curve. Nevertheless, we can say that the graph of the function represents the behavior of these data points well, showing dearly where the data rises and falls.

Guided Instruction

For this session segment, you will be working with the learner together. This is your chance to act as a guide for the learner and then allow him or her to use both their critical and creative thinking skills. In light of this, there are four teacher questions to help you with this process. Please be sure to use the teacher question answers to double check your work.

Question 1:

Consider the function $f(x) = 3\cos\left[\frac{1}{2}(x+30)\right] + 2$

Identify the transformations from the parent function $f(x) = \cos x$ and sketch the graph of the given function.

- What are the different stretch and shifts that happened in this function?

- How does the graph of this function compare with the parent function?

Question 1 Sample Student Response:

What are the different stretch and shifts that happened in this function?

This function had vertical and horizontal shifts, as well as horizontal and vertical stretches. The different transformations are :

$$a= 3 \quad b= \tfrac{1}{2} \quad c= -30 \text{ (30 units left)} \quad d= 2$$

How does the graph of this function compare with the parent function?

The transformed graph is wider and taller (stretched vertically) than the parent function. The graph also shifts up and left.

The graph of the function $f(x) = 3\cos\left[\frac{1}{2}(x+30)\right]+2$ is shown below:

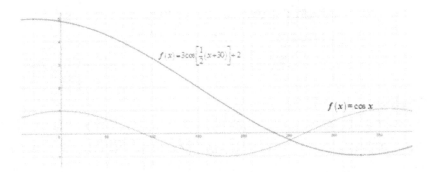

Question 2:

In Bayview Shore, high tides were noted at 1:05am and at 1:00pm while low tides were observed at 6:30am and again at 7:03pm. The following day, high and low tides were observed roughly at the same times as the previous day. Assuming that the high tide is estimated at 15 meters and low tide is at 3 meters, find a cosine function that best represents this phenomenon.

- How will you approach this problem?
- What are the amplitude, period, and principal axis?
- Is there a horizontal shift?

Question 2 Sample Student Response:

How will you approach this problem?

First, the data points will be plotted to visualize the behavior of the graph. Then the components of the function will be computed.

What are the amplitude, period, and principal axis?

Using the maximum value of 15m and minimum value of 3m, as well as the times when they occur; the above components are calculated as follows:

$$a = \frac{\text{max} - \text{min}}{2} = \frac{15 - 3}{2} = 6 \qquad d = \frac{\text{max} + \text{min}}{2} = \frac{15 + 3}{2} = 9$$

Since the firsthigh tide happened at1:05am and the next one at 1pm, the period can be estimated to be 12 hours.

$$b = \frac{2\pi}{12} = \frac{\pi}{6}$$

Is there a horizontal shift?

For a cosine function, the maximum value is on x = 0. In this scenario, the maximum value (ie. high tide) occurs at 1:05 am. If we are to use the x = 0 axis to be at 12:00am, we can say that there is a horizontal shift ofabout 1 hour (1 unit) to the right.

With the results found, we can represent the scenario using the following function in cosine:

Video Suggestions

While there are many videos available to help you, these are only to be used as a starting point to help you with any supplemental videos in which you may use. Please conduct a search on either YouTube or Teacher Tube to find appropriate videos for this lesson.

Below are some suggested title searches:

- Properties of Sine and Cosine Functions

- Transformations of Trig Functions-No Period Change

- Different Movements in a Carousel using Trig Functions

Independent Instruction: Working on Your Own

Now, it's the learner's turn. This is the learner's chance to demonstrate that he or she can complete the skill on his or her own. Depending on how much time in which you have left in the session, you may want to use only one or two of these questions and focus on the mini-assessment questions that are aligned with this particular lesson. Of course, it will depend on both the session time and the learner's progress.

Question 1:

Clearly draw the graph of the following function:

$$f(x) = -2sin\ (x + 90) + 3$$

Explanation (Sample response):

The transformations identified are:

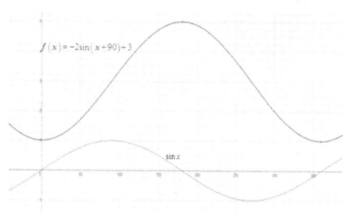

- **horizontal shift** of <u>90 units to the left</u>

- **vertical shift** of <u>3 units up</u>

- **amplitude** 2 with **reflection over the x-axis**

Question 2:

Find the function, in cosine, representing the graph below:

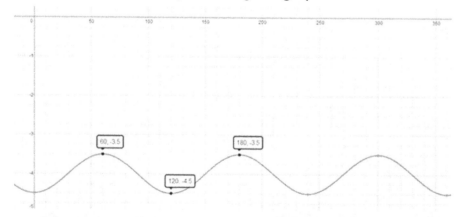

Question 2 - Explanation (Sample Response):

Looking at the graph, we can indentify the following points:
maximum: (60, -3.5) and (180, -3.5)
minimum: (120, -4.5)

Using these points, we can calculate the different components:

$$a = \frac{max - min}{2} = \frac{-3.5 - (-4.5)}{2} = \frac{1}{2} \qquad d = \frac{max + min}{2} = \frac{-3.5 + (-4.5)}{2} = -4$$

$$period = 180 - 60 = 120 \qquad b = \frac{360}{period} = \frac{360}{120} = 3$$

Using 360 in this formula as the unit is in degrees

Since the first maximum point occurs when x=60, the horizontal shift is then +60

Hence, the function representing the graph is:

$$f(x) = \frac{1}{2}\cos\left[3(x + 60)\right] + 3$$

Question 3:

The table below gives the average monthly minimum temperature, 0F, for Ross City in the polar regions:

Month	JAN	FEB	MAR	APR	MAY	JUN	JUL	AUG	SEP	OCT	NOV	DEC
Temp, °F	3.8	5.2	12.3	17.3	25.2	31.0	33.6	26.1	16.8	12.2	6.1	3.5

There is a function, $f(x) = A \sin [B (x-c)] + D$, that represents this data. Find **A, B, C,** and **D.**

Question 3 - Explanation (Sample Response):

First, let's converts the months into numerals: Jan=1, Feb=2, so on. Then let's first identify the maximum and minimum points:
maximum: (7, 33.6)   minimum: (12, 3.5)

Using the points identified, the following can be calculated:

$$A = \frac{max - min}{2} = \frac{33.6 - 3.5}{2} = 15.05 \qquad D = \frac{max + min}{2} = \frac{33.6 + 3.5}{2} = 18.55$$

The maximum and minimum points occur within 6 hours. This range of values is half the period. Hence, the period is 12 hours.

$$B = \frac{2\pi}{period} = \frac{2\pi}{12} = \frac{\pi}{6}$$

Since this is a sine function, and the first time the parent function has a maximum value is when x=π/2 (in radians since the x values are in decimal system), this function then moves to the right from x=π/2 to x=7. Hence, the horizontal shift is

$$C = 7 - \frac{\pi}{2} = \frac{14 - \pi}{2} \cong 5.4292$$

Mini-Assessment

At the end of your tutoring session, these are the questions that the learner will need to complete. Please make sure that you keep in mind only the topics in which you have covered in the lesson. As a suggestion, if time does not permit, you can have the learner complete the rest of the mini-assessment at the next tutoring session.

For questions 1-2, consider the function:

$$f(x) = 5 + 2\sin\left(\frac{\pi}{4}x\right)$$

1. What is the period of this function?

      a. 4

      b. 6

      c. 8

2. What is the maximum value of this function?

      a. 7

      b. 5

      c. 2

For questions 3-4, consider the graph:

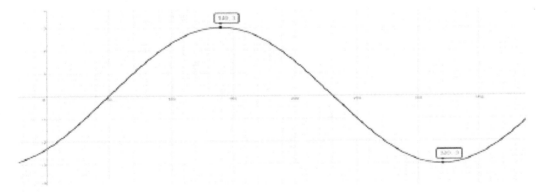

3. What is the period of this function?

      a. 180   b. 360   c. 540

4. Find the function, in sine, that represents this graph.

In an industrial town, the amount of pollution is greater during late afternoons with the factories in full swing, and lowers at midnight. With t being the number of hours after noontime, the amount of pollutants, in mg, per cubic meter of air is represented by

$$P(t) = 45 + \sin\left[\frac{\pi}{6}\left(t - \frac{37}{12}\right)\right]$$

5. What is the minimum level of pollution?

6. What is the maximum level of pollution?

7. At what time during the day does this maximum level occur?

Mini-Assessment Answers and Explanations

1. C

In this trigonometric function, $B = \dfrac{\pi}{4}$, Recall that the formula for B is $\dfrac{2\pi}{period}$;

hence, period $= \dfrac{2\pi}{B}$

$$\text{period} = \dfrac{2\pi}{\pi/4} = 2\pi \times \dfrac{\pi}{4} = 8$$

2. A

We know that the maximum value of *sine* is 1. The maximum value of the function also happens when sine is at maximum. Hence, the maximum value is **5 + 2(1) = 7**

3. B

From the graph, only the maximum and minimum points are given. For a trigonometric function, the period is twice the distance of maximum and minimum points. The graph's maximum point is at 140 and minimum is at 320. The difference is 180. Hence, the period is 360.

4. To find the sine function, we need to find the different components:

$$\text{amplitude} = \dfrac{max\text{-}min}{2} = \dfrac{3 - (\text{-}3)}{2} = \dfrac{6}{2} = 3$$

$$\text{principal axis} = \dfrac{max + min}{2} = \dfrac{3 + (\text{-}3)}{2} = \dfrac{0}{2} = 0$$

$$\text{for the period factor, } \boldsymbol{B} = \dfrac{2\pi}{period} = \dfrac{360}{360} = 1$$

Since we're looking for a sine function, let's recall where sine function "starts". The parent function has (0,0) as "starting" point. From the given graph, it looks like that starting point has moved to the right 50 units. HEnce, the sine function is:

$$y = 3 \; sin \; (x - 50)$$

5. The graph of the function looks like this:
Its maximum is at 46. Hence, the maximum level
of pollution is **46mg**.

6. The minimum point of the graph is at 44. So,
the minimum level of pollution is **44mg**.

7. Since we know that the maximum level is 46mg, we substitute is number to *P(t)*
and solve it for *t*.

*Given the solution on the right, t is **6 hours and 5 minutes**. This means the time
of day when the level of pollution is maximum is at **6:05pm**.*

$$P(t) = 45 + sin\left[\frac{\pi}{6}\left(t - \frac{37}{12}\right)\right]$$

$$46 = 45 + sin\left[\frac{\pi}{6}\left(t - \frac{37}{12}\right)\right]$$

$$1 = sin\left[\frac{\pi}{6}\left(t - \frac{37}{12}\right)\right]$$

$$sin^{-1}1 = \frac{\pi}{6}\left(t - \frac{37}{12}\right)$$

$$\frac{\pi}{2} = \frac{\pi}{6}\left(t - \frac{37}{12}\right)$$

$$3 = t - \frac{37}{12}$$

$$t = 6.0833$$

Lesson Reflection

In this lesson, we've learned:
- what a trigonometric function is and how it looks like.
- sine and cosine functions as the basic trigonometric functions.
- the different characteristics of a trig function and how to calculate for them.
- how to graph trig functions.
- the application of these functions in real world scenarios.

CPSIA information can be obtained
at www.ICGtesting.com
Printed in the USA
BVHW012340170920
589095BV00003B/4

9 781944 346614